DESIGNS IN SCIENCE
STRUCTURES

SALLY and ADRIAN MORGAN

Evans

EVANS BROTHERS LIMITED

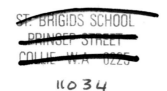
Evans Brothers Limited
2A Portman Mansions
Chiltern Street
London W1M 1LE

First published 1993

Printed in Hong Kong by Wing King Tong Co., Ltd.

ISBN 0 237 51259 9

Editor: Su Swallow
Designer: Neil Sayer
Production: Peter Thompson
Illustrations: Hardlines, Charlbury
David McAllister

Structures - Theory of

Acknowledgements

For permission to reproduce copyright material the authors and
publishers gratefully acknowledge the following:

Cover Adrian Morgan, Ecoscene
Title page Martin Bond, Science Photo Library **contents page**
Sally Morgan, Ecoscene **page 4** (top) Leeney, Ecoscene (bottom)
Robert Harding Picture Library **page 5** Bill Wood, Robert Harding
Picture Library **page 7** David Kay, Robert Harding Picture Library
page 8 (top) Robert Harding Picture Library (bottom) Professor
Harold Edgerton, Science Photo Library **page 11** (top) Robert
Harding Picture Library (bottom) Patrick Fagot, NHPA **page 12**
Manfred Kage, Science Photo Library **page 13** (top) Mary Evans
Picture Library (inset) Towse, Ecoscene **page 14** (top) Michael
Denance (bottom left) Philip Craven, Robert Harding Picture
Library (bottom right) Ken Griffiths, NHPA **page 15** (top) Steve
Robinson, NHPA (bottom) Hawkes, Ecoscene **page 17** (top) Sally
Morgan, Ecoscene (bottom) David Hughes, Robert Harding
Picture Library **page 18** (main picture) Stephen Dalton, NHPA
(inset) Groves, Ecoscene **page 19** (top) Glover, Ecoscene
(bottom) Chemical Design Ltd, Science Photo Library **page 20**
(top) Malcolm Fielding, The BOC Group Plc, Science Photo Library
(bottom) Jane Burton, Bruce Coleman Ltd **page 21** (left) Sally
Morgan, Ecoscene (right) Alfred Pasieka, Bruce Coleman Ltd

page 22 (top) Rolf Richardson, Robert Harding Picture Library
(bottom) Sally Morgan, Ecoscene **page 23** Winkley, Ecoscene
page 24 (left) Steve Alden, Bruce Coleman Ltd (right) Ecoscene
page 25 George Bernard, NHPA **page 26** Brian Hawkes, NHPA
page 27 (top) Dieter and Mary Plage, Bruce Coleman Ltd
(bottom) Sally Morgan, Ecoscene **page 28** (top) Rolf Richardson,
Robert Harding Picture Library (bottom) Jeff Foott, Bruce
Coleman Ltd **page 29** Ivan Polunin, NHPA **page 30** Walter
Rawlings, Robert Harding Picture Library **page 31** (left) Robert
Harding Picture Library (right) Christopher Rennie, Robert
Harding Picture Library **page 32** Otto Rogge, NHPA **page 33**
(top) Sally Morgan, Ecoscene (bottom) Peter Johnson, NHPA
page 34 (top) M. Prociv, NHPA (bottom) Anthony Bannister,
NHPA **page 35** Jones, Ecoscene **page 36** (top) Winkley, Ecoscene
(bottom) Hawkes, Ecoscene **page 37** Sally Morgan,
Ecoscene **page 38** Robert Harding Picture Library **page 39** (top
right) Werner Layer, Bruce Coleman Ltd (left) Simon Taylor,
Bruce Coleman Ltd **page 40** Dr Gopal Murti, Science Photo
Library **page 41** Sir Norman Foster and Partners **page 42** (left:
top, middle and bottom) John Shaw, NHPA (bottom right) Raynal
Saillet, Robert Harding Picture Library **page 43** (top) Shell (UK)
Picture Library (bottom) Emma Lee, Life File

Contents

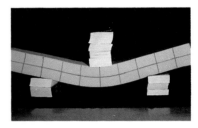

Introduction

An eggshell, a honeycomb, a bridge and a skyscraper are all structures. They can be quite simple, like the eggshell, or very complex, like the skyscraper. A structure usually consists of a number of smaller units joined together. Their shapes, and the way they are joined together, enable the structure to withstand a force, or even a number of different forces.

No matter how complex a structure may be, it must still obey the same physical laws as the very simplest of structures. For a structure to be successful, it must be able to carry the load for which it was designed, and withstand any external forces without collapsing.

Animals, plants and people build a vast array of structures of all sizes. Two of the largest structures in the world can actually be seen from the moon. The Great Wall of China was built by people, while the Great Barrier Reef, off the coast of Australia, was made by coral organisms. Both are made up of many smaller units. Even the tiniest creatures can create complex designs that rival those of modern architects.

A view over London docklands, showing the variety of structures that can be seen in any modern city

! *The Great Wall of China stretches 3500 km across northern China and is made from earth and stone. Its total length, including all branches, is nearly 10,000 km.*

The Great Wall of China

Measurement

These abbreviations are used in this book.

Units of length
km kilometre
m metre
cm centimetre
mm millimetre

Units of force
N newton
N/m² newtons per metre squared

Units of mass
g grammes
kg kilogrammes

Units of temperature
°C degrees Celsius

Units of area
ha hectare
cm² centimetre squared

Coral in close-up

Coral reefs are made by tiny animals called corals, which are related to sea anemones. Each coral has a soft body, known as the polyp, which secretes a limestone casing around itself. When the coral reaches its full size, it reproduces by sending out a thread which develops into another polyp. This grows and lays down its own limestone casing above that of the parent. As new individuals grow, the older ones are buried and die. The colony eventually consists of a thin outer layer of living individuals above many layers of empty chambers. While the bulk of the coral reef is dead, it still provides support for the living coral.

Materials and structures

Almost all of the materials used for making structures are solids, because solids are strong. Solid materials are made of many atoms, held together by strong chemical bonds. These bonds can be broken by force, by melting, or by chemical action. A piece of steel, for example, can be broken by bending it until it breaks, can be heated until it melts, or can be left in the rain until it rusts. The strength of a material comes from the ability of its bonds to resist external forces. The strength of a structure, however, depends not just on the strength of each material, but also on how the materials are arranged.

This book does not look at the materials themselves, but at the ways in which they are used by people, plants and animals to create structures. It examines the way structures are put together, the forces that act upon them and looks at structures of the future.

Important words are explained at the end of each section under the heading of **Key words** and in the glossary on page 44. You will find some amazing facts in each section, together with some experiments and some questions for you to think about.

Key words
Material any substance used to make a structure.
Structure an object that has a particular job and has to withstand forces.

The five main forces
The arrows indicate the direction of the force.

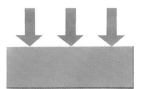

1 compression

2 tension

3 torsion

4 bending

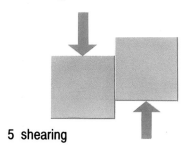

5 shearing

Forces in balance

Have you ever wondered why you do not fall through the floor when you walk about upstairs in a building? The simple answer is that the floor is strong enough to hold you up there. Engineers and architects have to make sure that floors can withstand the pushing forces, called compression, produced by the people and objects in the building. Compression is just one of five main forces that structures have to withstand. These main forces are :

1 A pushing force called compression

2 A pulling force called tension

3 A twisting force called torsion

4 A bending force

5 A shearing or tearing force

Sir Isaac Newton, an English scientist and mathematician, said that every force is matched and balanced by an equal and opposite force. When you stand on a floor, you push down on the floor with a force equal to your weight. At the same time, the floor will push up with a force equal to that of your weight. However, if the floor is not strong enough, then it will not be able to produce an equal and opposite push and you will fall through.

When a force is applied to a fixed object, it will cause the object to change shape. It is this change in shape, or deflection, that allows an object to resist the force. Although the object may not change shape significantly, some parts within it become squashed or compressed, while others become stretched. Every material allows a certain amount of movement. This is known as its flexibility.

When you walk across a plank of wood, it is deflected slightly – it bends. This produces sufficient internal forces to support your weight. How does this happen? When the plank is free from any load, the bonds between the atoms in the wood will be in a relaxed, unstressed state. The atoms always try to stay in this unstressed state. But if a load is put on top of the plank, the atoms immediately below the load are pushed together as the plank begins to bend. The wood on the underside of the plank is stretched or put under tension, and the atoms pulled further apart. Since the atoms always try to stay in the unstressed state, they try to return to their relaxed positions, producing the balancing forces needed to support the weight of the load.

When a load is put on a plank of wood, the wood on the bottom of the plank is stretched, while the wood on the top is compressed.

How strong is strong?

A strand of spider silk weighing 1 g has a greater tensile strength than the same weight of some kinds of steel.

Stress and strain

The words stress and strain are often used to mean the same thing. However, to an engineer they mean two very different things. Strain is a measure of how much the atoms and molecules of a material have been stretched apart. Stress is concerned with the force over an area of the material. If a 100 cm rod is stretched by 1 cm, it is subjected to a strain of 1/100 or 1 per cent. If the force causing the stretch was 100 N (newtons, units of force), and the cross-section of the rod was 100 cm^2, the stress would be 1 N/cm^2.

What features of metals make them useful when building structures?

Some of the latest man-made materials, such as fibreglass and carbon fibre, have higher tensile strengths than steel, but are still very light.

Canoes and kayaks are made of fibreglass or carbon fibre so they are light and strong.

The word strength is frequently misunderstood. It really refers to the force needed to break something. It must not be confused with stiffness, which is a measure of how flexible something is. For example, steel is stiff and strong but although a biscuit is also stiff, it is not very strong. The tensile strength of a material is the force needed to pull it apart by breaking all the bonds between the atoms. Many natural materials such as wood, flax and cotton have surprisingly high tensile strengths. Metals such as iron and steel have extremely high tensile strengths. They can withstand very heavy loads, so they are commonly used in large buildings and bridges. Metals can also be formed into complex shapes by hammering or by melting and then moulding. They remain strong, even in a completely new shape.

Building materials have to be able to resist both compression and tension. When a material is compressed, the bonds are squashed and the atoms pushed together. If the material is made of something soft, such as copper, it may be squashed out of shape. If it is brittle, then it may splinter and break. If the material is long and slender like a metal rod, it may buckle and either spring back into shape when the force is removed or stay deformed. Materials vary in their ability to resist forces. So designers choose different materials for different jobs in buildings. For example, bricks and blocks of stone are very strong in compression but cannot withstand tension nearly so well. In contrast, wood is stronger under tension than compression.

Bone is perhaps the most unusual structural material yet discovered, for it is almost as strong in compression as in tension. It can even repair itself when broken. It is found widely in the animal kingdom where it is used to give structural support to the largest animals (see page 39).

Concrete is poured over steel rods to strengthen it.

Some materials can be strengthened so that they can be used in a greater range of structures. Cement is made from limestone. When mixed with sand and water it forms a mortar which can be used to bind bricks and blocks together. If gravel is added, the mixture is called concrete, which is often used for the foundations and walls of buildings. Just like bricks and blocks, it is strong in compression and weak in tension. However, it can be reinforced so that it becomes stronger under tension. The concrete is poured over a lattice of steel rods. The rods help to prevent cracks from spreading through the concrete. To make the concrete stronger still, the steel rods can be put under tension. Small sections of concrete can be pre-stressed in a factory. The concrete is poured over rods that are held in tension. When the concrete sets, the tension on the rods is released, and the rods try to return to their original length. This compresses the concrete, making it stronger.

Plastic and elastic

If you squeeze a piece of Plasticine, clay or putty it will change shape but, when you let go, it does not return to its original shape. It has been deformed. Such materials are called plastics. The word plastic is often misused. The material we call plastic, used to make boxes, toys and even furniture, actually loses its plasticity during the manufacturing process. The hot, liquid plastic is forced into moulds and allowed to harden into the required shape.

Materials that can be stretched, squashed or bent but which return to their original shape are called elastic. Rubber, steel, and wood are all elastic, but only up to a point. If too much force is applied, they will not return to their original shape. Steel, for example, is quite elastic, but if a very large force is applied it will deform permanently. A

The pole vaulter's pole has to be flexible enough to carry the athlete over the bar.

Safety mats in sports halls and on the athletics field are made from foam. Why is foam a good choice of material?

steel coat hanger is a good example. If you squeeze it gently and let go, it springs back into its original shape. But if you bend it firmly, it stays deformed. If you keep bending it backwards and forwards, it will eventually break.

Surprisingly enough, glass is elastic. Thin glass fibres can be bent from side to side, and are particularly good at resisting tension. Glass fibres are used to reinforce poles for pole vaulting. They bend as the athlete jumps. But glass fibres are also brittle. If too much bending force is applied, they suddenly shatter.

EXPERIMENT

Elastic limits

This experiment tests the elasticity of different materials. The materials are stretched by hanging weights from them. You will need a selection of elastic bands of differing thickness and length, or some metal springs. You will also need some string or thin wire, two stools or chairs, a 60 cm length of wood, a long ruler and some weights ranging from 10 g to over 1 kg. You should wear goggles while you carry out this experiment.
1 Place two stools side by side with a 20 cm gap between them. Place the wood across the two stools. The wood will act as a firm support from which to hang the materials.
2 Measure the length of the material you are going test.
3 Hang the material from the wood, using the wire or string.
4 Attach a weight to the bottom of the test material, using string or wire. Measure the new length of the material.
5 Increase the weight on the material, noting the increases in its length. Repeat until the there is no more stretch.
6 Remove the weight.
Does the material return to its original length? If not, you have exceeded the material's elastic limit and it has been permanently deformed.
7 Repeat this experiment with the other elastic bands or springs. Do they behave in the same way? Is the elastic limit the same?

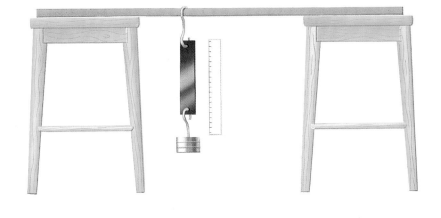

Key words
Compression a force that squashes an object.
Elastic describes a material that returns to its original shape after any stress is released.
Tension a force that stretches an object.
Torsion a force that twists an object.

Structural units

Most everyday structures, even those of the animal world, are made up from a number of smaller structures or building blocks. They are carefully joined together so that the whole structure can withstand a force. These building blocks include beams, pillars, domes and cables.

Beams and pillars

A beam is a long piece of wood, metal or other strong material used to span a gap and carry a load. It is designed to resist bending forces. A cantilever beam is one that is fixed at one end while the free end is able to support a load. If a beam breaks, it usually cracks at the point where it is under the greatest tension. For example, a cantilever beam will break at the point where it is fixed whereas a beam supported at each end usually cracks in the middle. Steel beams, often called girders, are not usually solid. In cross-section they are in the shape of a letter I or H. These shapes help to prevent them bending in one particular direction, and use less steel, which helps to keep their weight and cost down. Many bones act as beams, linking two parts of the body.

Pillars are long, just like beams. However, they are used in a vertical position and support a load acting downwards at the top. They are always in compression. Pillars have to be the strongest and thickest part of a structure, since they carry most of the load, so they usually have a large diameter. This makes them very heavy and, in very large buildings, much of the pillar's strength supports the weight of the pillar itself rather than the rest of the structure. The ancient Egyptians and Greeks made their temple pillars from stone. Each pillar was made from several solid blocks,

A beam spanning a gap, carrying a load in the middle. Where is the weakest point in this beam?

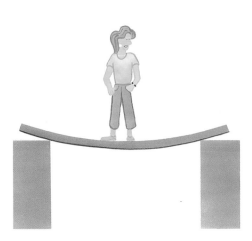

A cantilever beam, fixed at one end and carrying a load at the opposite end. Where is the weakest point in this beam?

each resting on the one below. Modern buildings often use reinforced concrete pillars, which are strong and cheap.

Pillars are found in animal skeletons, too, and they perform a similar function. In a four-legged vertebrate, such as the antelope, the leg bones act as pillars. Very heavy animals, like the rhino or elephant, have much thicker legs, for they have to support a much greater weight. It is very important that the pillar is not too heavy, or the animal will not be able to move its legs! (See below).

Pillars at Karnak in Egypt

The Hypostyle Hall at Karnak in Egypt has 134 pillars, each over 10 m in circumference and 50 m high. The roof has collapsed, but the pillars remain.

Hollow strength

Some of the smallest tubes ever made are used by doctors in microsurgery. They are 600 times thinner than a hair.

The hollow stems of some bamboos grow to 20 m or more.

It is often possible to improve the performance of a structure such as a beam or pillar by making it hollow. In a pillar, for example, the forces that resist compression and tension are stronger further away from the centre. The material in the middle of the pillar or beam adds little to the ability of the structure to resist bending, so it can be removed. Hollow beams and columns are almost as strong as solid ones but they are much lighter and cheaper to make because they use less material. During the construction of a building, a strong and lightweight frame of hollow steel tubes, called scaffolding, is used to support the workers and their materials. Nature has evolved design solutions that also make use of the advantages of hollow tubes. Many plants have hollow stems with bundles of supporting tissue arranged in a ring towards the outside of the stem. This arrangement makes the stem light and strong. Bamboo is a very good example. Its hollow stems are extremely resistant to compression and bending forces. In the Far East, where bamboo grows freely, it is often used in scaffolding since it is very cheap, is widely available and is easier to handle than steel.

Many of the bones in a vertebrate skeleton are hollow. A long bone such as the femur (the thigh bone), would be far too heavy to move if it were solid. Instead, it is a hollow tube that is both light and strong. This shape is more efficient for a bone than the I or H of a steel girder (see page 10), because it resists force equally well in all directions. The femur has a thin, outer layer of solid, very compact bone. Inside, there is spongy bone, which is made of a

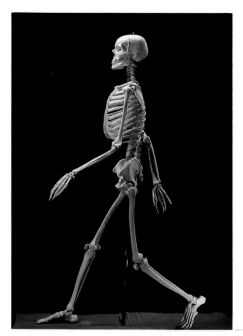

The longest bones of the human skeleton are hollow, so they can be moved easily.

? *Can you think of any other hollow structures? What are they used for?*

lattice of fine bone running through spaces filled with blood and jelly-like marrow. The fine bone is laid down in the direction of the main lines of stress and tension, for extra strength. The outer layer is quite thin over the ends of the bone, where it is subjected to less force, but is thicker in the middle, making it stronger and more resistant to bending and torsion.

For animals that fly, weight is critical. A bird weighs very little, and so its long wing bones have little weight to carry but have to withstand quite large bending and twisting forces. The bones in an albatross wing have almost reached perfection. They are hollow, thin-walled cylinders, in a superb combination of strength and lightness.

Propeller shafts in large ships are round and hollow, too. They have to withstand torsion as the blades of the propeller are twisted through the water. Torsion is strongest on the outside of the shaft, so a hollow shaft places the steel in the best position to resist torsion.

There are many advantages in using hollow structures but there is a limit to how thin the walls of a hollow structure can become. If they are too thin, small sideways forces would cause them to crumple. For example, the amount of force required to crush a drinks can is very large if it is pushed from both ends. But a much smaller force is required to crush the sides. This design may be ideal for carrying a liquid but it would not be suitable for a larger structure.

Roofs, leaves and other sheets

A sheet is a flat piece of material. People, plants and animals all make use of sheets. For example, leaves, bat wings and floors are all sheets. Sheets are usually supported at the edge, and carry a load spread over the surface of the sheet. Because they are quite large, it is important that they are light as well.

Plant leaves are crisscrossed by a network of veins. These veins have two functions. As well as carrying materials to and from the cells within the leaf, they also provide structural support. The central, and usually the largest, vein is called the midrib. The huge round leaves of the royal water lily of the Amazon can grow up to 2 m in diameter. They are thin yet strong, and the ribs help them to keep their shape. The ribs radiate like the spokes of a wheel, from a central hub. They are deeper at the centre and flatter near the edge. The ribs also divide and inter-link to form a network of short beams to hold the whole structure together.

Some architects have based designs for buildings on the structure of this water lily. Joseph Paxton studied the plant before he designed the Crystal Palace, built to house the Great Exhibition in London, in 1850. The Crystal Palace was built of glass and iron, on a plan of radiating ribs and cross-ribs. Ridges and furrows in the roof directed the rainwater down gutters into hollow pillars.

! *The leaf of the Amazonian Bamboo Palm can reach almost 20 m in length and over 1 m in width.*

It took 83,600 m^2 of glass to cover the Crystal Palace. Six million people visited the exhibition during the five months that it was open. Then the whole structure was dismantled and moved to another site in London. It burnt down in 1936.

Crystal Palace (above) and the royal water lily (right) both use a radial pattern of ribs.

The leaves of the royal water lily are so strong that a small child can stand on one.

Similar designs can be seen in the roofs of some sports stadia, where the supports radiate from a central point.

The wing of a bat is made from skin stretched over a supporting framework of bones. The thin sheet of skin is very light, having no fur covering, yet it is flexible and tough enough to withstand the stresses of flight.

Sheet materials made by people tend to be much stiffer than those found in the living world. Flat roofs, for example, are often made from sheets of plywood protected from the weather by sheets of tar-coated felt. The plywood itself is made up from a number of separate thinner sheets of wood. The wood grain in adjacent layers runs in different directions, to make its strength the same at any angle.

A sheet of flimsy material can be made quite stiff by folding it a number of times. The folds help to stop the material buckling as a load is applied. Palm trees grow in regions with storms and high winds, so the leaves have to be strong enough to withstand very bad weather. Some palms have leaves that are folded into ridges which make the leaf stiff and hard to tear apart. Corrugated roofs use the same design technique to turn a thin metal or plastic sheet into a rigid, waterproof shield.

How many different materials in sheet form can you find in and around your home?

Shells and domes

The shell shapes that sit on the cultural centre in New Caledonia cast shade and direct cool winds down into the building.

The igloo is a strong structure. What makes it strong?

In the natural and man-made worlds, shells and domes are used for protection. They are strong and often lightweight structures. The curved designs spread any load over a large area. Sometimes the shell or dome has ridges or thickened rib sections that provide extra support and rigidity, just like the ribs of the water lily. (See page 13).

Many soft-bodied animals have shells to protect them. Bivalves such as scallops and cockles have two hinged shells.

A cathedral roof and a skull are both dome-shaped. For buildings such as cathedrals and sports stadia, domes are an ideal way of covering a large area uninterrupted by supporting pillars. Tortoises, turtles and limpets all have dome-shaped shells which protect their bodies. An eggshell, which is really two domes joined together, protects the developing embryo. A surprising amount of force is required to break an egg by squashing the ends together.

The Pacific Rim Cultural Centre, New Caledonia is dominated by tall shell shapes. Each shell has inward-curving ribs of pine joined by stainless steel rods. These shapes not only provide protection from the weather, but also help to ventilate the building.

Buildings and animals both use domes. The dome of this Florence cathedral (left) spans a large space. Tortoises (below) can pull in their head and legs under their domed shell for protection.

Triangles and hexagons

How could you make a four-sided structure rigid?

The triangle, with three sides, and the hexagon, with six, are important shapes in both natural and man-made structures.

The triangle is one of the most basic engineering shapes. Three beams can be joined together to form a strong, rigid frame. Large structures are often made up from many triangular frames, joined to one another. For example, electricity pylons are constructed from many steel girders connected in triangles. The triangular design allows rigid, three-dimensional structures to be built using the minimum amount of material. Loads are spread through the structure, with some beams being in tension and others under compression. If the joints are firm, the structure will be strong and rigid. This type of construction is also found in cranes, bridges, and tent frames.

Many seed pods are triangular in cross section, providing a strong protective case for the seeds, yet remaining light in weight. Even giraffes use the triangle made by their forelegs as a stable support when they splay their legs in order to reach drinking water.

Honeycombs are based on the hexagon. Honey bees build a wax comb of hexagonal cells with walls that meet at exactly 120°. This precision is no accident. It allows the maximum number of cells to be fitted into the least space. There is no wasted space between the cells. The walls

When a giraffe bends over to drink it becomes unstable. So it spreads its front legs to form a stable triangle.

A pylon is made up of many triangles forming a strong, stable shape.

Can you think of five other structures that are based on the triangle?

are designed to be only as thick as stresses on the structure require, and no thicker, so they can hold the most honey for the least wax. A bee measures the thickness of the cell wall by pressing on it with its jaws and seeing how much the wall bends. Charles Darwin studied honey bees closely. He said of the honey bee: 'Beyond this stage of perfection in architecture natural selection could not lead; for the comb of the honey bee, as far as we can see, is absolutely perfect in economising labour and wax.'

Plant leaves need to be strong, yet light enough for the plant to be able to hold them out in the sun. When viewed under a microscope, the pattern of cells in the leaf between the upper and lower leaf surfaces resembles a honeycomb.

Man-made structures also use the honeycomb. Aircraft bulkheads and floors, for example, have to be strong and rigid, yet very light. Honeycombs of metal are sandwiched between thin sheets, making a material with just these characteristics.

! *Honey bees build their combs with astounding precision. The thickness of the walls varies by less than 0.002 mm.*

EXPERIMENT

Testing shapes

In this experiment you will test the strength of different shapes. You will need some pieces of thin card, each 20 cm x 15 cm, sticky tape, and some small books to act as weights.

1 Make the following shapes with your card, using sticky tape to hold them in place: a round tube, a triangular tube, a rectangular tube and square-sided tube.

2 Test the strength of these shapes by laying the tubes on their side and placing a book on each one. You may have to lean the book against some shapes. Gradually increase the weight of books on the shapes.

Which shape was the strongest? What happens if you stand the tubes on their end rather than on their side and put the books on top?

EXPERIMENT

Hexagonal shapes

This experiment investigates the pattern of closely-packed bubbles. You will need a clear plastic bottle with a screw cap, some soapy water, some narrow rubber tubing, a pair of scissors and sticky tape.

1 Using the points of the scissors, carefully make a hole in the side of the bottle just large enough to push the rubber tubing through. Seal the point of entry of the tubing with the sticky tape.

2 Pour a small amount of soapy water into the bottle and tightly screw the cap in place.

3 Blow bubbles into the solution using the rubber tubing. Make enough bubbles to fill the bottle.

4 Look closely at the shapes of the bubbles. The bubbles in the middle will form hexagonal shapes rather like a honeycomb.

Chains and cables

How many different types of chains and cables do you use each day?

A cable car suspended from a thick cable

Chains and cables are both composed of strands of material. They have thousands of uses for people, from shoelaces to washing lines, sewing thread to anchor chains and tow ropes to bridge supports. They are used to attach two objects to one another. A piece of string or a chain is always used under tension. They are no use at all in compression, for they can only withstand a force when pulled. If the tension is released, they simply fall in a heap.

Huge steel cables are used in some of the world's longest bridges, in mountain cable cars and are also used to support very tall structures such as radio masts. They support a bridge by holding it up, and a mast by pulling it down, but they are always under tension.

Cables under tension are used by living organisms as well. The roots of a plant act as cables, anchoring the plant to the ground. The bones and muscles of a mammal are joined together by cables called tendons and ligaments. Tendons join muscle to bone. They have to be tough and inextensible so that they do not stretch when the muscles contract. Ligaments, on the other hand, join bone to bone. They have to allow enough movement for the limb to bend, and so they are slightly elastic.

A spider's web is a structure that relies entirely on the properties of cables for its strength. The cables are made from a protein-

! *The strongest wire rope ever made is 282 mm in diameter, and can withstand a force of 3250 tonnes.*

! *The largest web made by an orb web spider is 1.5 m in diameter and has anchor lines 6 m long. The silken cables are so strong that they can be used as fishing lines.*

rich substance like silk secreted by the spider. Spiders' webs seem to be very fragile structures, but the silk is surprisingly strong. It is also elastic, so that it can absorb the impact of a flying insect.

The way the spider constructs this web is a good example of how a structure composed entirely of cables can be built. The female orb spider starts to spin her web by making a small fan-shaped kite which she allows to be carried off on the breeze, making sure a thread connects the kite to her body. When the kite snags on a twig or branch she anchors the other end, and begins the process of web-building. She walks along the thread and, halfway across, she drops down to the ground in order to secure another thread to act as an anchor. This forms the basic framework. More radiating lines are added before she starts to build the spiral threads from the centre outwards.

The spiral part of the web is actually built twice. The first time it is built from non-sticky silk. Then she rebuilds the outer part of the spiral, which will trap her insect prey, by eating the non-sticky silk and replacing it with sticky threads. The spider's feet are coated with oil so that she does not stick to the web herself.

Some modern buildings, such as those at Expo 92 in Spain, use cables to support parts of the structure. Avenues are shaded by sheets of material that are hung from a network of cables. Temporary shelters such as tents use cables to hold them in place.

The cables supporting this tent-like roof (inset) are attached to a mast in the same way as this spider's cobweb (main picture) is anchored to a plant.

Spirals

Can you find other examples of spirals indoors and outside?

Spirals are common in the natural world, but less so in man-made structures. A spiral, like the hexagon, is an important shape because it makes good use of space with a minimal amount of material. A coiled telephone cable can be stretched, but is very compact when not in use. A spiral staircase takes up less space than a straight flight of stairs.

Spirals are particularly important to plants and animals. For some, they allow growth to take place easily and without any change of shape. Snail shells have a spiral shape. As the snail gets larger, it simply grows a new twist to the spiral. For others, the spiral is used to save space. Flower petals are often arranged in a compact spiral inside the bud which protects the petals before the flower opens. Sunflower seeds are arranged in a tight spiral, making the best use of the space available for seed storage.

There is so much genetic information in cells that it must be packed into the nucleus as efficiently as possible. DNA, which is short for deoxyribonucleic acid, is the molecule that contains this genetic information, passing it from generation to generation. DNA controls every aspect of development in the body. It is made up of two molecular chains wound together in a double spiral called a helix, just like a twisted ladder. The two chains form the sides of the ladder and the rungs are made from cross-links between the two chains. The double spiral is a wonderfully efficient way of packing the enormous amount of information stored in DNA into the confined space of the nucleus of a cell.

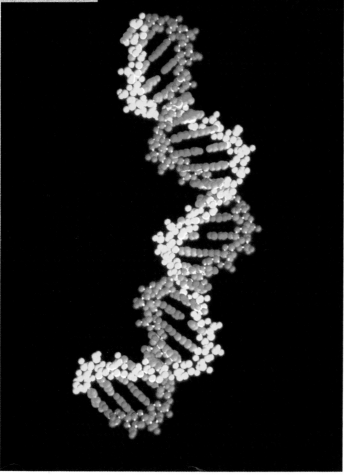

The spiral is a compact shape used in man-made and natural structures. Both the helter skelter (above) and DNA (right, a computer image) use a double spiral.

Key words
Beam a long piece of material that is put across a gap and designed to carry a load.
Cantilever a beam that is fixed at one end and carries a load at the other.
Dome an arched shape that spans a large space.
Pillar a vertical beam.
Sheet a flat piece of material.

Frames and skeletons

In many modern glass-walled buildings, the support comes from the framework on the inside, not from the glass on the outside.

? *Some of the skull bones of a newborn baby are not fused together. They fuse during the first few months of life. Why are they not fused at the time of birth?*

A close-up of the skeleton of the Venus flower basket sponge

Many structures rely on a framework for support. Animal skeletons are frameworks that support and protect the organs of the body. Such skeletons have some structural parts, such as the skull and rib cage, and joints between bones to allow movement. Plant skeletons support the stem and leaves. Buildings, too, need a frame on which the materials that make up the structure can be supported. Many modern buildings have a framework of steel bars, while others have supporting walls of brick and concrete.

Skeletons are multi-purpose devices. Not only do they provide support and protection, but they also allow a wide range of movement. They form a framework to which the muscles can be attached and against which they can act. There are three main types of animal skeleton. In endoskeletons, the bones are found inside the muscles. Vertebrates such as fish, reptiles, birds and mammals all have an endoskeleton. Animals such as insects have an exoskeleton, which forms a tough outer covering of the body with the muscles inside. Worms have hydroskeletons: their support comes from a fluid within the body.

Endoskeletons are made up of a series of bones linked together by joints, with the muscles outside the bones. Bones are moved as muscles contract. Some of the bones, such as the skull and hips, are made of several bones fused together. This gives extra strength and support.

Exoskeletons are found in a number of animal groups, the most common being the arthropods. These are animals that have jointed limbs, such as crabs, spiders, centipedes, millipedes and all insects. An external skeleton is both supportive and highly protective, but it does have some disadvantages. A hard exoskeleton cannot expand to allow growth, unlike bone that can keep increasing in size until the animal is fully grown. This means that, in order to grow, animals with an exoskeleton have to moult, shedding their old skeleton to reveal a new one underneath. Before the new skeleton hardens, the animal expands by taking in fluids. Many insects have to moult several times before they reach full size. Exoskeletons are quite heavy, and make movement difficult. This means that animals with exoskeletons cannot grow very large unless, like lobsters and crabs, they are supported by water.

Sponges are marine animals that have an endoskeleton. The Venus flower basket has a tubular, lace-like endoskeleton made of spines of a hard material called silica. The spines interlink to form a framework. This framework, which is covered by a layer of living tissue, is a highly efficient design. It provides great strength with the minimum amount of material.

The human skeleton contains more than 200 bones. There are 27 bones in the hand alone. The smallest bones are in the ear. One, the stirrup, is just 3 mm long but it is essential for hearing.

Plants have an internal skeleton. All plant cells are surrounded by a wall made of cellulose. This gives the cells protection and support. Running throughout the plant are a series of tubes called xylem vessels that transport water while at the same time providing support. The plant's skeleton cannot be seen from the outside while the plant is alive, but is often revealed when a plant dies because the strengthening material takes a long time to decay.

There are many similarities between skeletons and buildings. The outer walls of a traditional building provide support in the same way as an exoskeleton. But the structure of some modern buildings is based on a frame which acts like an endoskeleton in vertebrates. Glass-walled buildings, for example, have an internal framework of steel or concrete. The outer glass wall does not provide any support; it is just a protective shield, rather like a skin.

In the deserts of Arizona, USA, the skeletons of the huge saguaro cactus are left standing for many years after the cactus dies.

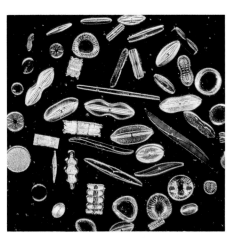

Small plankton called diatoms make surprisingly strong structures. These tiny plants have a silica skeleton made in two halves called valves. The valves overlap, forming a structure that resembles a box with a lid.

Key words
Endoskeleton an internal skeleton, with muscles on the outside.
Exoskeleton an external skeleton with muscles on the inside.
Skeleton the framework of an animal or plant.

The box girder is a type of beam bridge that is extra strong because of the use of triangles. This one is being built in Thailand.

Bridges, dams and tunnels

Bridges, dams and tunnels are all similar because they all span a gap and carry a heavy load. Many use an arch somewhere in their design. All these structures are found in the man-made and the natural world.

It has taken people thousands of years to develop the skills needed to build these structures, which are often complex. Many of the solutions have been found only by trial and error. The same design challenges have been faced by animals. They have evolved their own solutions over an even longer period.

Beam bridges

The first bridges were probably logs laid across streams. Today, very long bridges over rivers, roads and railways are vital parts of our transport systems. The main problem with bridges is that they tend to sag in the middle, so a key part of bridge design is concerned with finding ways to prevent this. There are three basic types of bridge, known as beam, arch and suspension.

The simplest design of bridge is a load-bearing beam that is supported at either end. It is suitable for bridges that do not need

EXPERIMENT

Examining beam bridges

In this simple experiment you will discover which part of a beam bridge is compressed and which part is stretched. You will need a piece of firm foam 1 m long x 10 cm wide x 10 cm deep, a felt-tip pen, a long ruler and some books.

1 Draw a series of vertical lines at 10 cm intervals along the length of the foam.

2 Draw a horizontal line across the middle of the vertical lines. You should now have two rows of rectangular shapes.

3 Make two piles of books of equal height and lay the foam across the gap. To start with, put the book piles close together. Place one book in the middle of the foam to represent a load on a bridge. Add more books one by one until the bridge sags.

What happens to the foam?
Which parts of the beam have been compressed? Which have been put under tension?
Repeat this experiment, placing the piles of books further and further apart. How many books do you need each time as the load, to make the bridge sag?

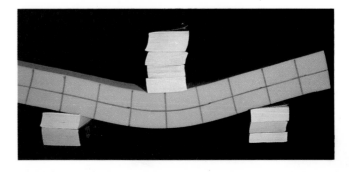

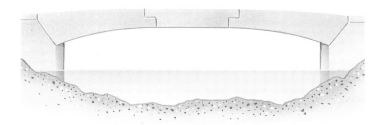

This simple cantilever bridge uses two fixed beams, with a third beam in the middle.

to be very long or high. A plank across a stream is a beam bridge. Usually the ends of the beam rest on supports. It can be strengthened by putting one or more pillars, called piers, under the middle of the span.

A cantilever bridge uses beams to provide support and allow a greater span. Two outer beams, one on each side of the gap, hold up a third beam in the middle. One end of each outer beam is held firmly down, while the middle of each outer beam is often supported by a pier. The free end of each outer beam is therefore capable of supporting the central beam.

Arch bridges

Arches are capable of spanning large openings where an ordinary beam may collapse. A stone arch is made up of a number of small, wedge-shaped pieces of solid material which stay in place because their own weight causes them to press evenly against one another. The central wedge, or keystone, transfers its load to its neighbours, which in turn pass their load on down the sides of the arch. Modern arches are often made in a single piece, by pouring concrete into a mould. In arch bridges, the load tends to push the piers outwards, so massive end piers and foundations are often needed for support. Sometimes, in river bridges, a whole series of arches is necessary for the load to be transferred to the bank.

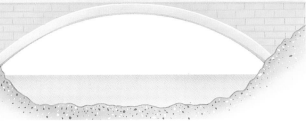

This arch bridge transfers the load it carries to the river banks.

This aqueduct in Spain, once used to carry water, is made up of two rows of arches that span a large gap.

The earliest known arches were used in Mesopotamia about 4000 years ago. The Romans developed the arch design further, and made some spectacular aqueducts with multiple-arched bridges on top of one another.

Many structures, other than bridges, make use of arches. Where else have you seen arches used?

Natural arches are found in many rock formations. They are a result of the action of weathering of the rock. Erosion by water and wind causes the rock to be slowly worn away. Often the arch shape is all that remains, because this shape is a very strong, stable one.

Suspension bridges

Suspension bridges span large gaps with ease. The main support comes from a pair of huge cables that are held high above the road by pillars at either end of the bridge. The cables are not drawn tight, but drape across the gap. Lighter, vertical cables reach down from the main cables to support the roadway. The cables are all under tension, even though the draped ones look as if they are sagging. The pillars are under compression. Suspension bridges are relatively simple and cheap to construct and make efficient use of materials.

Suspension bridges are used all over the world. The Golden Gate bridge in San Francisco, USA, (main picture) is made of steel. Elsewhere, bamboo or other natural materials can be used instead (inset, in Papua New Guinea).

The world's longest single-span bridge is being built in Japan, to a cable-stayed design. It will join the islands of Honshu and Shikoku. When finished in 1998, the Akashi Kaikyo Bridge will be 3560 m in overall length, with a central span of 1990 m.

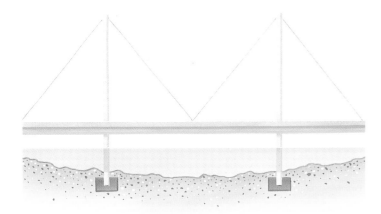

A recent modification of the suspension bridge is the cable-stayed bridge. The supporting cables reach directly from the roadway to the top of the nearest pillar. This design simplifies the construction and further reduces the materials required, making it the most efficient design yet developed for large-span bridges.

The cable-stayed bridge is like a suspension bridge but simpler.

An animal bridge

Can you see any similarities between the rabbit skeleton on this page and a bridge?

The skeletons of many four-legged animals are often compared to a bridge. When the animal is standing square on four legs, most of its weight is suspended from the backbone. The backbone is made from a series of smaller bones, called vertebrae, each separated from the next by a small disc of cartilage. The backbone is usually under compression, rather like the roadway of a bridge. Ligaments also join vertebra to vertebra. Being under tension, the ligaments strengthen the backbone and prevent too much movement. Most of the weight is taken by the back legs and hips in the same way that piers take the weight of a bridge. The weight of the head is supported by the bones of the shoulder. Muscles and ligaments hold it steady, regardless of the other movements of the body. With all these bones, ligaments, and muscles, the animal backbone is much more complex than a simple man-made bridge. Not only must it be able to support the weight of the body, but it must also be able to bend from side to side and flex up and down. Although all bridges sway, the degree of flexibility found in the animal backbone would be the last thing an engineer wants in a bridge.

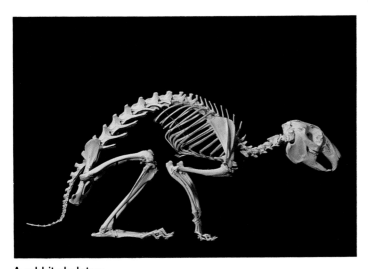

A rabbit skeleton

Challenge! Build a spaghetti bridge

In this challenge you have to design a bridge to carry model cars across a gap at least 30 cm wide. The bridge must be capable of carrying two model cars placed side by side in the middle. All you have to complete this challenge is a pack of long spaghetti and some water!

Clue The spaghetti does not have to be dry and it can be broken into shorter lengths. You could even cook it.

Dam builders

A dam is a barrier across a stream or river. It stops the flow of water, which collects behind the dam and forms a large pond or lake. Dams are made by animals as well as people.

The beaver, found in North America and parts of Europe, is an expert dam builder. A dam helps to protect a beaver from predators, and provides a home as well as a storage place for food.

Beavers create the main dam wall by ramming sticks upright into the stream bed and then placing small trees and branches across them. The whole structure is weighed down with small boulders. Mud is used to bind all the parts together and make the dam watertight. The upstream side is made steep and is well covered with mud, whilst the downstream side slopes gently and is covered with poles laid parallel to the stream bank. This helps the dam withstand pressure from the newly-formed lake. A spillway at each end of the dam allows excess water to escape. The dam requires constant maintenance. The spillways have to be deepened after heavy rain to relieve the pressure on the dam and then have to be rebuilt later, in order to stop too much water leaving the lake. Beaver dams may last many years and serve several generations of beavers. These dams often change the appearance of a whole valley, forming a series of small lakes where once there were none, creating new wetland habitats. Eventually the lakes silt up and turn into swamps to be colonised by other plant species. The beavers are forced to leave as the water disappears, and dry grassland eventually develops.

People build dams too. Some are built to prevent flooding by controlling the flow in a river. Some provide water for irrigation schemes, while others use the water to generate electricity.

The largest type of man-made dam is the gravity or embankment dam. It is broadest at the base, where the pressure from the water is greatest, and slopes up to a relatively narrow ridge along the top. The central core of clay or concrete extends deep into the earth, to prevent water seeping underneath. A concrete dam may have hollow spaces inside to reduce the amount of material needed for construction. The spaces also allow maintenance teams to inspect the dam.

Less bulky dams include the cantilever dam, which is reinforced with a web of steel bars, and the arch dam which, in common with arch bridges, transfers the weight of the water it supports to the sides of the valley.

The world's largest dam is at Itaipu on the borders of Brazil and Paraguay in South America. It is 8 km long, 180 m high and contains 28 million tonnes of concrete. The Chinese plan to build

The beaver has very strong teeth to gnaw through small trees for building dams.

A beaver dam of mud and sticks can have the tensile strength of reinforced concrete.

How can water be used to generate electricity?

This dam on the Colorado River in the USA has to be strong enough to support the weight of the water in the lake.

an even larger dam across the Yangtse river. The Three Gorges Dam will produce 84 billion kilowatt-hours of electricity each year from hydro-electric power, equivalent to one-sixth of the country's total energy generation. More than 1.7 million people will have to be resettled to make way for the rising water levels.

EXPERIMENT

Water pressure

This experiment shows how water pressure increases with depth of water. You will need a watertight plastic bottle, a pair of scissors, some sticky tape and some water. You should do this experiment on the draining board near a sink, or outside.

1 Cut the top off the plastic bottle.

2 Use the point of the scissors to make four small holes, equally spaced from the top to the bottom of the bottle. Make sure you take great care when using the scissors.

3 Cover all the holes with a single long piece of sticky tape.

4 Fill the bottle to the brim with water.

5 Quickly remove the tape and watch how the water flows through the holes.

Where is the strongest jet of water? Does this explain why a dam is thicker at the bottom than at the top? Why is no water flowing out of the top two holes in the bottle in the photograph?

Tunnelling

 The latest tunnel-boring machines can dig out a tunnel 3 m in diameter and 114 m long in just one day.

People and animals build tunnels and they both have to overcome similar design problems. Tunnels must be kept dry, they must not collapse and the air inside must be kept fresh.

Modern technology allows engineers to build tunnels through most types of ground, even through ground that may seem unsuitable because it is not firm, or is waterlogged. The tunnel may be lined, to provide extra strength or to prevent the tunnel from flooding. To prevent the ground from collapsing while the tunnel is being constructed, engineers use a tunnel shield, a metal cylinder that is moved along the tunnel as digging takes place. Behind the shield, the tunnel is lined with either concrete or metal. Tunnels need to be ventilated, to keep the air fresh. Engineers solve the problem by siting large fans at the top of ventilation shafts to suck stale air out of the tunnel.

Animals have evolved specially adapted feet and teeth for digging. Most animals that dig tunnels are found on grassland where the ground is easy to burrow through. Their tunnels are usually built with circular or arched sections, which give the maximum strength to resist the pressure of the earth above. Because animal tunnels are usually small, they do not often require special linings or ventilation systems.

A huge tunnel-boring machine used to dig the Channel Tunnel between France and England, the world's longest rail tunnel

Prairie dog

The naked mole-rat spends all its life underground in a maze of tunnels that act as living quarters, nurseries and larders. These mole-rats have become well adapted to living underground. They have lost their hair and eyes, but their incisor teeth are very much enlarged, forming an ideal tool for digging. They avoid swallowing the soil by keeping their lips tightly closed around their incisors. Mole-rats work as a team to build their network of tunnels. The first in line removes the soil and pushes it back to the next, who pushes it further back, and so on until the last mole-rat in line tosses the soil out of the burrow. In the sandy soils of East Africa where they live, they can dig incredibly fast – faster than any predator that might want to dig them up.

Prairie dogs, which build quite long tunnels, have solved the problem of ventilation. Their tunnels have an entrance at each end. One end opens flat on the ground while the other forms a

Prairie dogs burrow a maze of tunnels, called a city. The largest city ever recorded was thought to house over 400 million prairie dogs!

raised chimney some 30 cm above the ground. Since wind moves faster a little way above the ground, the air passing across the entrance of the chimney is moving faster than the air at the other end. Stale air is drawn out of the raised end of the tunnel while fresh air is sucked in at the other end.

Many spiders build tunnels. The purse web spider has a silk-lined tunnel about 45 cm long which is sited near the base of a tree and is camouflaged with leaves. When an insect walks over the tunnel, vibrations are set up which alert the spider, which then bites through the tunnel and grabs its victim.

The female trap-door spider is also a tunneller. She makes a short burrow about 10-15 cm into soft ground and then lines her trap with silk. She also makes a perfectly camouflaged, circular lid by binding together tiny soil particles with silk. She gives the lid a silken hinge and attaches gravel to the underside so that it will close under its own weight. When a small animal such as an insect passes by, she grabs it and pulls it into her trap. The door automatically shuts behind her so the prey cannot escape.

A trap-door spider at the entrance to its tunnel

The longest tunnel in the world is the Delaware aqueduct in the Catskill Mountains, USA. It is 169 km long, and carries water to New York City.

*Miners used to light fires at the bottom of mine shafts to ventilate the tunnels. How did these fires bring about ventilation? (For a clue see page 32 **Building a home**.)*

Key words
Arch a curved shape used to support a bridge, or the side of a building.
Suspension when a load is supported from above by cables.
Dam a barrier that stops the flow of water downstream.
Spillway the overflow channel at the side of a dam that allows excess water to escape.

Building design

It is obviously very important to get the design of a building right, for it must be strong enough to withstand all the forces to which it will be subjected. As well as supporting the weight of the building itself and the additional loads it carries when in use, the design must allow for other factors, such as winds and earthquakes. The same principles of design apply to buildings made by people, homes made by animals and even to many large plant structures.

Foundations

The deepest roots, produced by the wild fig in the semi-desert of southern Africa, can reach a depth of 100 m below ground level.

All tall structures need good foundations to support their weight and to resist sideways forces that might topple them over. Foundations are just as important in the natural world as they are in man-made buildings. The most familiar natural foundations, which have many parallels in human design, are the roots of trees.

Trees are the tallest living plants and it is easy to imagine that their roots, which help to support the tree and act as anchors in the ground, go deep into the ground. However, strong roots do not always have to be deep. Where solid rock lies only a metre or so below the surface of the soil, tree roots fan out over the rock, forming a root plate several metres across. Some root plates can extend as far along the ground as the tree is high.

Some trees, found mainly in the rainforests, have developed extensions to their trunks, called buttresses. It was assumed that these buttresses helped to support the tree in the thin soil of the forest, but recent research has indicated that trees with buttresses are no more successful than trees without them. Their real purpose remains a mystery, but it may be to do with competing for nutrients. However, the buttresses found on the outside of large buildings such as cathedrals do have an important supporting function.

Buildings need foundations too. Older buildings had foundations only a few feet deep, often because the importance of good foundations was not understood. Modern house foundations are now at least 1 m deep, and considerably more in soft soil. In some areas the whole house is built on a reinforced concrete raft, to prevent subsidence.

Foundations for houses are usually based on trenches filled with concrete. For larger buildings, long piles of steel or reinforced concrete are driven into the ground. The piles are overlaid with concrete to form the base of

Most tree roots are quite shallow but spread out over a wide area.

The foundations of the Leaning Tower of Pisa in Italy (above) were not deep enough and the tower started to lean. Modern towers (left, the World Trade Centre in New York) usually have very deep foundations to support them.

The Leaning Tower of Pisa is tilting at a rate of 1.02 mm per year.

the building. If there is hard rock near the surface, foundations may only be a metre or two deep. Many of the taller New York skyscrapers on Manhattan Island have quite shallow foundations as they are built on rock. The tallest buildings in New York are the twin towers of the World Trade Centre at 415 m high. Their foundations had to be much deeper than those of any other Manhattan skyscraper because they stand on soft ground beside the Hudson River. The earth was dug out to a depth of six storeys and filled with concrete. A dam had to be built around the excavation site to stop the river water pouring in. London stands on clay, so buildings there also need very deep foundations.

The Leaning Tower of Pisa in Italy is a very good example of what happens when foundations give way. The tower was built in three phases, starting in 1173, and started to lean immediately. It tilted first one way and then, as the second phase started, it tilted in another direction. When the seventh storey was completed, the tower suddenly began to tilt to the south. The final storey was built at an angle to the rest of the tower, to try to make it appear as if the tower was vertical. Each year the tower tilts slightly more. The tilt is caused by an unstable layer of clay beneath the tower, the layer on the south side being thinner than that to the north. The tower now leans at such an angle that it is unsafe, and it is propped up by cables and counterweights. One interesting proposal to restore the tower to a vertical position involves a process called electro-osmosis. An electric current would be passed through the soil on the north side to extract the water and thus cause the soil to shrink.

Building a home

! *If there was a human equivalent to a termite nest, it would stretch over 2 km into the sky.*

Animals are very skilled at making homes. There are animal examples of stone masons, weavers, plasterers, miners and sculptors. Many animals, such as spiders, swifts and bees, can actually make their homes out of secretions from their own body.

African termites produce vast, complex cities using mud glued with saliva. A complete colony consists of many millions of termites and the nest itself can rise several metres above ground.

Air conditioning is achieved through a remarkable natural design. Approximately 2 m below the surface is a huge, uninhabited chamber. At the base of this chamber are shafts that descend a further 4 m or more. A massive clay pillar in the middle of the lower chamber supports the ceiling and the bulk of the nest. Thin vertical vanes hang from the roof of the chamber. The mud in the vanes absorbs moisture from the nest above. As the water evaporates it uses up heat energy, which in turn cools the air. This makes the chamber the coolest place in the nest. Warm air, from the main part of the nest, rises by convection to the top of the towers, where it leaves through the mud walls. Passageways near the surface of the nest draw cool, oxygenated air down into the underground chamber, where it is cooled further. This cool, fresh air then rises through the nest. This feat of engineering is even more impressive when it is realised that the massive structure is built entirely by teams of blind workers in total darkness.

Termite nests are very tall structures built by tiny animals. The nests also extend several metres down into the ground.

Investigating convection

Many ventilation systems rely on the principle of warm air rising and cooler air being drawn in to replace it. Warm air is less dense and so lighter than cold air, so it will rise up. The same applies to warm and cold water. You will need two large plastic bottles, two 10 cm lengths of plastic tubing, a nail, Plasticine, crystals of potassium permanganate (from a chemist's), two dishes and plenty of ice.

1 Cut off the tops of the plastic bottles.

2 Use the point of the scissors or a nail to make two holes on one side of each bottle, one hole 2 cm from the top and one 4 cm from the bottom.

3 Place the bottles 8 cm apart, holes facing each other, and insert the plastic tubing in the holes to form two bridges, one at the top and one at the bottom. Seal the joins with Plasticine.

4 Stand each bottle in a dish. Pack ice around the base of one bottle and pour warm water around the base of the other.

5 Carefully pour water into the bottles. When everything is ready, drop a few crystals of potassium permanganate into the bottle

standing in warm water. They should sink to the bottom and start to dissolve. Watch what happens to the purple dye.

Explanation The warm water around the first bottle warms up the water inside while the ice cools the water in the second bottle. The warm water will move through the tubing into the top of the cold water bottle while cold water will move across to the bottom of the warm bottle. The potassium permanganate helps you see this happening.

The huge cooling towers seen around power stations work by convection too. The hot water produced by the waste heat energy from the power station warms the air in the tower. As the warm air rises up the chimney, it is replaced by cool air being drawn in at the bottom.

It is not only termites that make elaborate homes. Green tree ants of Australia make a nest from the leaves of the tree in which they live. The larvae of the ant produce silk threads that are used to stick leaves together. Slowly the ants link a number of leaves together, forming a conical nest within the tree.

Weaver birds weave their own nest in the same way as a person would weave a piece of cloth. The bird starts its nest by knotting a long piece of grass around a branch. Then it threads strips of grass across other threads that run at right angles to each other. The finished nest dangles from the branch. Weaver birds can weave dome-shaped nests, and ones with several compartments. Some even construct waterproof roofs.

The wren-tit builds an elaborate nest, using sticky cobwebs to bind the twigs together. First, a thick foundation is formed as the bird weaves a network of cobweb across the twigs. Then the sides are built up, using tiny fibres of bark held in place by more web.

The weaver bird weaves a nest from strands of plant material.

Most bird's nests have a similar shape. What shape is this? Why is it a good choice?

Why is it important to have the right amount of moisture in clay?

These nests of the white-rumped swiftlet are on the ceiling of a cave in Australia. They are made of grass, twigs and saliva.

Swifts, too, are expert nest-builders. Remarkably, they manage to build a nest on a cliff face or side of a building without landing, for their legs are not designed to support the weight of their bodies. The cave swiftlets of southeast Asia build a nest using their own saliva, which they produce in large amounts. The swift chooses a nest site on the cliff and uses its long tongue to dab saliva on to the rock as it flies past. The bird makes hundreds of passes of the nest site, each time adding a dribble of saliva. The saliva quickly hardens and gradually the wall of the nest takes shape. Once the nest is large enough, the bird finishes the nest from the inside, forming the cup shape common to all swift nests.

A potter wasp lays her egg in the finished chamber made from clay.

Clay is a very versatile material that is used by people and animals alike. It can be moulded while it is still wet and forms a tough structure when it dries and hardens. The female potter wasp is good at using clay to make a safe chamber in which to lay her single egg. The clay has to contain exactly the right amount of moisture, so if it is too dry she will add water to it. She rolls the clay into a strip using her jaws and front legs, and lays it in a ring on the building site. She returns with further strips of clay that are added to the existing rings, slowly building up the chamber. Once the chamber is large enough, she searches for a suitable prey which she stuns with her sting. The prey is placed in the chamber along with one egg and the chamber sealed. This way the wasp ensures that the larva will have plenty of food when it hatches!

Man-made buildings

In the past, houses were built with walls made of a single thickness of material. More recently, houses have been built with a double skin of material, separated by a cavity. The cavities are either hollow, or are filled with an insulating foam or fibre. The two skins are held together with wall ties. The outer skin, which is exposed to the weather, may be brick or concrete, while the inner skin can be made from lighter materials; often these are aerated concrete blocks or sometimes just timber and plasterboard. Buildings with more than a single storey obviously need extra floors. In a two-storey house, a number of wooden beams, called joists, are placed from one inner wall to another and are covered by floorboards. Different construction methods are used in multistorey buildings. In the unit slab method, each floor is built as a unit on top of the floor beneath, using steel or reinforced concrete beams to support the weight of the higher floors. In the frame method, a complete steel frame is erected and the walls and floors of the building are constructed later. The cantilever method uses one or more large pillars to act as supports for floors and beams. This method was used to build the 47-storey Hong Kong and Shanghai Bank building in Hong Kong. It has eight pillars standing on a concrete pier that reaches 35 m into the rock below ground. The floors are suspended from five huge horizontal beams which are clearly visible from outside, giving the building its distinctive appearance.

Very tall buildings have to withstand the force of high winds. Those in earthquake-prone areas also have to be able to withstand earth tremors. Modern skyscrapers are designed to be flexible so they can absorb much of the force of the wind or earthquake without collapsing.

The design of the Hong Kong and Shanghai Bank makes a feature of some of the structural supports.

What is the function of the cavity in a cavity wall?

However, if these buildings were to sway too much, the people inside would feel very uncomfortable. One solution to this problem is to position huge weights on rollers at the top of the building. As the building moves, the weights shift position and damp the motion. The Citicorp Building in New York has a huge concrete damper weighing 400 tonnes at the level of the 59th floor. It is connected to the framework by shock-absorbing arms and floats on a film of oil. When it is windy, more oil is pumped under the damper, lifting it up and allowing it to move, and its slow movements counteract the sway caused by the wind. Counterbalancing weights are also used in the towers of the largest suspension bridges for the same purpose.

The CN Tower in Toronto, Canada, at 555 m, is the world's tallest self-supporting structure. There are taller structures, but they are supported by cables. Although made of steel and concrete, it is not as rigid as it looks, for in high winds the top can sway up to 0.5 m from the vertical.

Tall trees

The shapes of broadleaved trees are easy to see when they have lost their leaves in winter.

The tallest things in the natural world are trees. The problems faced by tall trees are the same as those faced by buildings. Trees, too, have to be able to sway and bend in the wind. Trees are large, and need a lot of food, so they have a lot of leaves for photosynthesis. Their leaves give them a very large surface area, so in windy weather they suffer from large sideways forces as the leaves catch the wind. This is particularly true for broadleaved trees which have larger leaves than the conifers. Broadleaved trees require strong trunks for support and broad roots for good foundations. They tend not to grow quite as tall as conifers, but spread their branches more widely. They lose their leaves in autumn, so the wind resistance is much less during the winter months when the weather is more severe. Conifers, on the other hand, tend to have a triangular shape that is superbly designed to withstand wind and shed snow. They never lose all their leaves at once, so wind

EXPERIMENT

Tree design

This experiment investigates the flexibility of different lengths of wood of the same thickness. The wood will represent the tree trunk. The longer the piece of wood, the longer the tree trunk. You will discover whether taller or shorter trees have more flexibility. You will need a piece of wood dowelling 60 cm long and 1 cm in diameter, a 1 kg weight, a clamp stand and boss, and a ruler.

1 Clamp the piece of dowelling at 90° to the stand, using the clamp and boss. Start off with a long piece of wood by adjusting the dowelling so that you have a 50 cm length. Attach the weight 5 cm from the free end of the dowelling. If the weight slips you may have to make a small notch in the wood.

2 Measure the amount by which the dowelling bends.

3 Repeat this again but adjust the dowelling so that the trunk is only 40 cm long, then 30, 20, and 10 cm long. Use the same weight each time. Which length of wood was the most flexible? How does this relate to the height of different trees near your home?

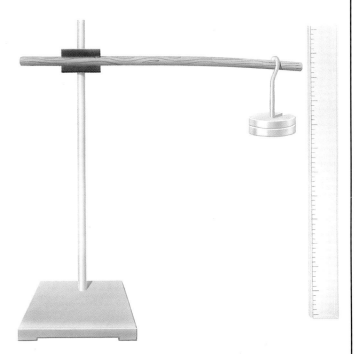

When there is a constant wind coming from one direction trees grow away from the wind to reduce the force of the wind on its trunk and branches.

resistance is the same all year round. Trees can also respond to external forces in other ways. Trees that grow where there is a prevailing wind, for example, lay down extra wood to help resist the sideways force.

 Have a look at the trees growing near your home. What types of trees are the tallest? Which have the greatest girth (circumference)? Can you see a relationship between height and girth?

Key words
Convection the rising of warm air or fluid because it is lighter than the cooler surroundings.
Foundations the parts of a structure that are below ground, supporting the parts of the structure that are above ground.

Problems with being big

The weak pull of gravity on the moon allows astronauts to take larger steps than on earth.

There are particular problems that are associated with being very large. There are large animals, plants and buildings, but is there an upper limit to their size? Are there any factors that might prevent an animal on land becoming three times the size of an elephant? It turns out that there are some very important physical factors that affect size. These are gravity, kinetic energy and surface tension.

Gravity is a force that affects all structures, whether living or non-living, small or large. Gravity acts on an object by pulling it down towards earth. However, smaller structures are less affected by gravity and the smallest insects almost seem to behave as if gravity did not exist – some of them jump to many times their own height. It is gravity that makes the body's mass act like a weight and the force exerted by gravity depends on the size of the object. The greater the mass of an object, the more it is affected by gravity. If the force of gravity were to double, we would not be able to walk upright, and most animals would have evolved short legs, perhaps like those of a crocodile. On the other hand, if gravity was halved then we would have lighter, more slender limbs. The moon has only one sixth the gravity of earth, so astronauts are able to move around with ease, taking giant steps.

Kinetic energy is the energy that something has because it is moving. A moving car has kinetic energy. So does a falling object. It is kinetic energy that causes our bodies to be injured when we fall. This energy is transferred to the ground on impact. A child half the size of an adult will hit the ground with only one-sixteenth of the force that would be experienced by the adult. This is because the child only weighs one quarter of the adult's weight. Small animals are little affected by kinetic energy but this force does affect the design of larger animals. The larger the animal the heavier its skeleton must be to allow it to withstand falls.

Surface tension is a molecular force that pulls the surface of any liquid into the minimum possible area. Water molecules are strongly attracted to each other and the bonds that form between them are difficult to break, so the water holds together. A tight skin forms on the surface. It is this property of water that can cause difficulty for small animals. When a small fly falls into a pool of water, for example, it is trapped by surface tension. Surface tension does not affect larger organisms such as ourselves for its force is weak in comparison to the strength of our muscles.

! *Raindrops can be a major hazard to insects. On an equivalent scale, it would be as though you were hit by a huge cannon ball of water.*

Gibbons have slender bones which allow them to swing easily through the trees. But their bones are not strong: one gibbon in three will break a bone at some time in its life.

The gazelle (top right) is slender with long legs. The elephant (above) is much heavier, so its legs are much shorter and thicker.

The bones of larger animals must be thicker and shorter to support the extra weight, and so the whole skeleton is bulkier and heavier. For example, bones make up 8 per cent of the weight of a wren or mouse, 14 per cent of a goose or dog, 18 per cent of a human and 27 per cent of an elephant. Since larger animals are heavier, they tend to be less agile. Although this might not seem to be the case for the giraffe, since they can move with great speed, they lead a risky life. Their bones are long, relatively light, and so are easy to move. But they are not strong and, should the giraffe lose its balance and fall, its bones shatter quite easily.

Elephants are not at the upper limit of size; in the past there have been much larger animals. The brachiosaur, a type of dinosaur, was 12 m high and weighed 80 tonnes, 16 times the weight of an elephant. Scientists have calculated that the theoretical maximum weight of a skeleton made of bone could be as much as 100 tonnes. So why are there no truly giant animals today? It is all a question of time! Large animals live a long time. Their hearts beat slower than small animals and they produce fewer young. They take a long time to reach maturity and even their period of pregnancy is long. The gestation period for an elephant is 22 months. Large animals need stable environments in which to live. These species evolve slowly and they cannot respond to climate changes as rapidly as a small animal species with a

Viruses are 50 times smaller than bacteria. They can only be seen with an electron microscope, magnified more than 30,000 times .

The virus is the smallest known organism. These have been magnified 36,000 times.

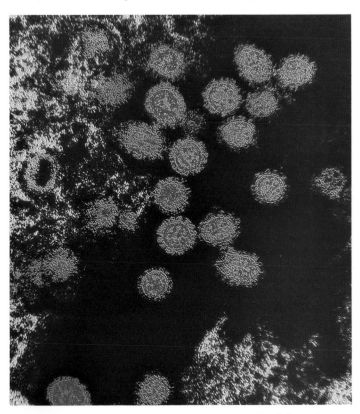

If humans used up energy as quickly as the shrew they would have to eat 30 kg of food every day.

Why do whales stranded on beaches usually die, even if they are helped back into the water?

rapid reproduction rate. The life span of an elephant may be as long as 70 years. In one elephant's lifetime there may be over 50 generations of a small mammal such as a mouse. If the environment changes rapidly, the larger species tend to die out, since they cannot adapt quickly enough. It is therefore time that limits the maximum size of animal species.

What, then, limits the minimum size of an organism? The smallest mammal is the shrew. It has to eat for much of the day, just to provide enough energy to keep warm. In fact it has to eat at least half its weight in food each day. If it were any smaller, it would have to eat all day long. Animals are made of many millions of cells, the basic building blocks of an organism. But there is a limit to how small the cell can become before it will cease to work properly. The very smallest living organisms of all are called viruses. They can barely be seen with the most powerful microscopes and their structures are very different to those of other organisms, for they are not composed of cells at all. A virus has a very simple structure consisting of a spiral of genetic material protected by a spiral of protein. They cannot respire, excrete or grow, causing many scientists to question whether they are, in fact, alive. They can only reproduce inside living cells of other organisms, so they are parasites. Once a virus has invaded a host cell, it takes over control, instructing the cell to make more copies of the virus.

None of the really large animals have exoskeletons, which suggests that there is some limiting factor affecting the size of such a skeleton. It turns out that the reason is to do with the strength of curved plates. As the length and breadth of a curved plate increases, it cannot support the same weight unless it is made much thicker. This creates a weight penalty. It is only at sea, where water provides most of the support, that animals with exoskeletons, such as crabs, can grow to any real size.

In the same way that the size of animals is limited by physics, the size of man-made structures is similarly limited. Very large buildings exert tremendous stress on their foundations. If there is too much stress, the bottom of a building will be unable to support its weight. Miniaturisation – making things very small – also poses problems. Microchip circuits, for example, are very familiar nowadays, and are continually getting smaller. However, it is becoming increasingly difficult to make such circuits even smaller. This is partly because very special equipment is needed to operate at such fine detail. It is also because the tiny currents of electricity flowing within the circuits actually behave differently when flowing through very tiny structures.

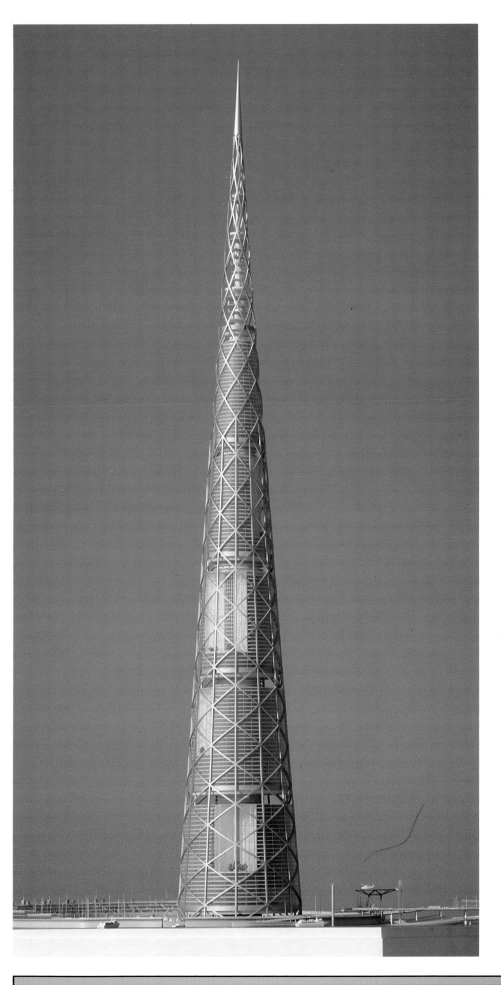

Area and volume

Growth in length and increase in volume are related. It is obvious that a large animal is heavier than a small one, but a small increase in size actually produces a large increase in mass. Imagine a cube 1 cm x 1 cm x 1 cm (volume of 1 cm^3) and a mass of 1 g. Now if the object doubles in size to 2 cm x 2 cm x 2 cm, its volume increases to 8 cm^3 and its mass becomes 8 g. So, doubling the length increases the weight 8 times.

If you have read the story of Gulliver, you can understand how worried the Lilliputians must have been. Even though he was only 12 times the height of a Lilliputian, his body would have had a volume equal to 1728 (12x12x12) of theirs. In practical terms, he would have needed 1728 times as much food as one of them!

The Millenium Tower could be built in Japan in the 21st century. It would be twice as high as any building in the world today. But will technology allow us to build such a tall structure?

Key words
Gravity a force that draws two objects together.
Kinetic energy the energy possessed by a moving object.
Surface tension molecular force that pulls the surface of any liquid into the minimum possible area.

caterpillar

pupa

adult butterfly

Some animals, like this monarch butterfly, can reorganise their structure and alter their appearance completely. Scientists may find clues in the natural world that will help them design structures that can change shape.

The future

If we can be certain of anything, it is that the future of structural design will be exciting. New design techniques and understanding of materials will make structures more efficient and safer.

Advances are being made in the creation of 'intelligent structures', structures that can detect and respond to forces. Engineers have traditionally thought of structures as passive, fixed things only affected by external forces. Biologists, on the other hand, study a world in which things grow, reproduce and move of their own accord. These living organisms respond to forces, and can show signs of damage. For example, if you bang your leg, you feel pain and a bruise appears. Engineers are trying to produce similar responses from man-made structures. Climbers may soon be using ropes that change colour when the rope is subjected to excessive strain. The colour change will tell climbers whether the rope is safe to use. A plastic bag handle may change colour when it is overloaded and about to break. It could be that in future a building might contain a control and communication system linked to its structure, warning the engineer if any part of the building was damaged.

Trees and bones respond to stress by laying down more material where it is needed to counteract the stress. A 'smart' timber bridge could, perhaps, gradually change shape to spread the load more evenly. Some materials can now respond to heat or force by changing colour. Coating a building in a colour-sensitive material, which turns dark in low temperatures and lighter in higher ones, would help the building warm up quickly yet stay cool in the heat of the day.

New materials and design techniques are helping engineers produce more adventurous structures. Some of these are in hostile

Soon, climbers may be using ropes that change colour if they are damaged.

A design for a gas drilling rig of the future. It is superimposed on a photograph of a city to show the size of the structure that would sit out at sea. It is possible to build such a tall structure at sea because the water will help to support it.

environments such as space, where both NASA of the USA and the European Space Agency have advanced plans for permanently-manned space stations in orbit around the earth. The oil industry is continually finding new deposits of offshore oil, but in ever deeper water. Huge new drilling rigs are being designed using new techniques of construction and the most modern materials, to safely operate in some of the most dangerous seas in the world. Oil companies have plans to explore for oil in some of the deeper oceans, to depths of 900 m. Shell are currently building a new drilling platform consisting of a steel floating deck moored by steel cables anchored to the sea bed. It will pump oil from 32 wells. The whole structure will be 1007 m tall, twice the height of the world's tallest building, the Sears Tower in Chicago. Even larger rigs are planned, with two decks and 64 wells, having the ability to drill for oil in over 3000 m of water.

Inventive designers are able to combine new materials with old ideas to come up with exciting new structures. The ancient Japanese art of paper folding, origami, has many techniques for turning a single sheet of paper into a complex shape. The special folds have been applied to modern plastic sheet materials, the end result being a new kind of tent that packs flat, yet can be easily unfolded. Because some of the folds create ridges, the design is extremely sturdy. The tent is ideal for use in disaster relief operations, for it is simple to erect, easy to transport and cheap.

Many of the structures with which we come into everyday contact will benefit from such technological advances and many of these advances will come about through study of the natural designs found in the animal kingdom.

Origami techniques are being used to design new, easy-to-build shelters.

Glossary

aqueduct a bridge that carries water.

arthropod an animal that has an exoskeleton and jointed limbs, such as insects and crabs.

buttress a support on the outside of a wall.

cantilever a beam that is fixed at one end and carries a load at the other.

cell the building block or basic unit of an organism.

convection the rising of air or a fluid because it is lighter than the cooler surroundings.

dome a hemisphere shape that spans a large space.

energy the ability to do work.

evolution a very slow process of change that affects all living organisms.

endoskeleton an internal skeleton, with muscles on the outside.

exoskeleton an external skeleton with muscles on the inside.

force something that changes the shape or motion of an object.

foundations the parts of a structure that are below ground, supporting the parts of the structure that are above ground.

genetic code information that is required to control the cells of an organism.

girder a large steel beam.

gravity a force that draws two objects together.

kinetic energy the energy possessed by a moving object.

nucleus the control centre of a cell that contains the chromosomes.

nutrients any substance that feeds an animal or plant.

parasite an organism living in or on another organism, from which it obtains its food.

photosynthesis the process by which green plants make their own food using light energy, carbon dioxide and water.

predator any carnivorous animal.

silica a common, very hard material found in shells and rocks.

spillway the overflow channel at the side of a dam that allows water to escape.

strength a measure of the force required to break an object.

tissue a group of similar cells with a particular function, for example liver tissue or muscle tissue.

ventilation a system that allows fresh air to pass through a space.

vertebrate animal that has a backbone.

Index